Work 243

菌菇网

The Internet of Fungus

Gunter Pauli

[比] 冈特·鲍利 著

[哥伦] 凯瑟琳娜·巴赫 绘

朱　溪 译

上海远东出版社

丛书编委会

主　任：贾　峰

副主任：何家振　闫世东　郑立明

委　员：李原原　祝真旭　牛玲娟　梁雅丽　任泽林

　　　　王　岢　陈　卫　郑循如　吴建民　彭　勇

　　　　王梦雨　戴　虹　靳增江　孟　蝶　崔晓晓

特别感谢以下热心人士对童书工作的支持：

匡志强　方　芳　宋小华　解　东　厉　云　李　婧

刘　丹　熊彩虹　罗淑怡　旷　婉　杨　荣　刘学振

何圣霖　王必斗　潘林平　熊志强　廖清州　谭燕宁

王　征　白　纯　张林霞　寿颖慧　罗　佳　傅　俊

胡海朋　白永喆　韦小宏　李　杰　欧　亮

目录

Contents

一些松甲虫准备向长满桦树和冷杉的森林发起进攻。他们朝着树木飞去，开始初尝树的滋味，却很快意识到味道不佳，所以决定离开。

"这些树怎么会知道我们要来呢？"年轻的松甲虫想。

Some pine beetles are ready to launch an attack on a forest full of birch and fir trees. They move towards the trees and take their first few bites but quickly realise it just does not taste good, so decide to move on.

"How did these trees know that we were coming?" the young beetle wonders.

一些松甲虫准备发起进攻……

Some pine beetles are ready to launch an attack …

... a traitor amongst us ...

"我们当中一定出了叛徒，"年长的松甲虫回答道，"这些树一定是提前得到了消息，才会发出这么可怕的味道！我们要好好查一查。"

　　"但不可能是我们内部的问题。我们又不会和树木说话。他们的语言实在不同，我都不知道从何学起。"

"We must have a traitor amongst us," an older beetle responds. "These trees must have had advance notice for them to have developed such a horrible taste! We need to investigate this."

"It cannot possibly be any of our family members. We do not know how to talk to trees. Their language is so different, I wouldn't even know where to start."

"那肯定又是这些蘑菇作祟。"

"蘑菇？可是他们都不能和树说话啊？"

"哎，其实他们会的。他们说的语言所有人都能明白——人人为我，我为人人。当一切结束，每个人都开开心心，对彼此充满信心——多少年来都如此。"

"It must have been those mushrooms again."

"Mushrooms? But surely none of them can speak to trees?"

"Oh, but they do. They speak a language that everyone understands: I do this for you and you do that for me. And when it is done, everyone is happy and have confidence in each other – for years to come!"

……他们还会合作。

... they are cooperating.

"照你这么说，好像他们交流很多，可是我却从来没听到过他们之间交流过一个字。"

"他们不仅会沟通，还会合作。"

"很显然他们能和所有的物种说话。我们刚到，这片区域里所有树木的味道就变得如此糟糕，以至于我胃口全无。"

"You make it sound as if there's a lot of communication, but I have never heard one word exchanged."

"They are not only communicating, they are cooperating."

"And they can apparently talk to everyone because when we arrived, all trees in this area were tasting so bad that I completely lost my appetite."

"瞧，菌菇在土壤里探索，先找到食物和水分，再供给树木，然后换来糖分。"

　　"蘑菇没有牙齿，所以也用担心吃糖吃坏了牙齿吧，我想。"

"Look, the fungus explores the soil, finds food and water, gives it to the tree – and gets sugar in return."
"Mushrooms don't have teeth, so do not need to worry about the sugar, I suppose."

……换来糖分。

...and gets sugar in return.

他们可是网络大师……

They are masters of the network ...

"他们可是网络大师，可以在黑暗里工作，只有在子实体准备好之后才会突然冒出来。他们还会不断与冷杉、枫树、雪松和桦树交换数据。"

　　"菌菇还能和草说话，是真的吗？"年轻的松甲虫问道。

"They are masters of the network. They operate in the dark, only popping up into the light when their fruit bodies are ready. And they continuously exchange data with firs, maples, cedars and birches."

"Is it true that fungi can also talk to grass?" the young beetle asks.

"的确如此！他们可以和所有有根系的物种讲话。在仅仅一平方米的范围里，有长达一千多千米的菌菇'电线'连接着每种活着的植物！"

"因此，与拥有手机网络甚至互联网的人们相比，他们的联系更加便利？"

"Indeed! They are able to talk to anything that has roots. In just one square metre of space, there are more than a thousand kilometres of fungi 'wires' connecting every living plant!"

"So they are better connected than people with their cell phone networks – or even their internet?"

......可以和所有有根系的物种讲话。

... able to talk to anything that has roots.

......即使最小的树也有着和最大的树同样可怕的味道。

... even the smallest trees had the same horrible taste.

"唯一的区别在于他们的通信不是无线的。他们之间的直接联系比人类想象的还要多。"

"你有没有注意到，即使最小的树也有着和最大的树同样可怕的味道，极其强烈，极其令人讨厌？"

"当然注意到了，这是母亲在照顾自己的后代。"

"The only difference is that their communication is not wireless. They have more direct connections than people can even start to imagine."

"And did you notice that even the smallest trees had the same horrible taste as that largest tree with the strongest, most repugnant taste?"

"Of course, that is a mother taking care of her offspring."

"但是母树怎么认识她那从种子长大的后代呢？她怎么与他们直接交谈的呢？"

"嗯，如果阿里巴巴或亚马逊通过网络能知道孩子们在哪里以及他们在做什么，那为什么一棵树就会不知道？这完全归功于菌菇网络的互联世界啊！"

……这仅仅是开始！……

"But how does the mother tree know that her offspring came from her seed? And how is it that she can talk to them directly?"

"Well, if Alibaba or Amazon knows through the network where people's kids are and what they are doing, why would a tree not know? It is all thanks to the interconnected world of the Internet of Fungus!"

... AND IT HAS ONLY JUST BEGUN!...

AND IT HAS ONLY JUST BEGUN! ..

There are mother trees in the forest. These trees take care of the younger ones. When information arrives through the fungal internet, the mother tree will send carbon to an injured sapling, and distribute warning signals.

森林里也有母树。母树会照看年幼的树。当信息通过菌菇网到达时，母树将向受伤的树苗运送碳，并发出警告。

As the climate changes, native tree species are replaced by new tree species. The mother tree sends nutrients, guides the new neighbouring seedlings to give a head start to assume a dominant role in the ecosystem.

由于气候变化，本土的树种会被新树种取代。母树运送养分，引导邻近新的幼苗开始在生态系统中发挥主导作用。

The language of communication amongst trees by way of fungi is not just resource transfer. The trees have the ability to call for defence and to recognise kin. The behaviour of the plants is constantly adapted.

树木之间通过菌菇进行语言交流不仅仅是为了资源转移。树木有寻求保护及认识亲属的本领。植物的行为会不断调整以适应环境。

In areas devastated by the pine beetle, it was found that fungal diversity and the mycorrhizal network help regenerate the new seedlings that were coming up in the understory, by guiding their defence mechanism.

在遭受松甲虫破坏的地区，人们发现菌菇依靠自身的多样性和菌根网络来指导植被的防御机制，从而帮助植被新苗再生。

$\mathcal{A}$ Douglas Fir forest regenerates biodiversity by way of the fungal network that increases the productivity of generating biomass, from improving carbon cycling to the increase in bird and insect diversity.

道格拉斯冷杉森林通过菌菇网再生了生物多样性，从改善碳循环到增加鸟类和昆虫的多样性，该网络提高了生物量的生产能力。

$\mathcal{C}$limate change, pine beetle infestations, and logging disrupt vital communication networks in the forest. Loggers have become sensitive to the threats to big game and wildlife, but never considered the mycorrhizal fungi.

气候变化、松甲虫侵扰和伐木活动破坏了森林中重要的通信网络。伐木工人已经很容易对大型动物和野生动植物产生威胁，但还从未考虑过菌根真菌。

An old tree stump, one that foresters would consider without life, actively connects and communicates with all plants species around it, through the fungal internet. There is no quick end to the life of a tree.

一个古老的、被认为是没有生命的树桩，会通过菌菇网络积极地与周围的所有植物物种建立联系并进行通信。一棵树的生命不会迅速终结。

The fungal network is an information superhighway, speeding up interactions between a large, diverse population of plants. This internet sabotages unwelcome plants by spreading toxic chemicals.

菌菇网是一条信息高速公路，可加快大量不同植物种群之间的相互作用。这一网络会通过传播有毒化学物质来破坏不受欢迎的植物生长。

Is there really a biological internet?

真的有生物网络吗?

Would you prefer to be connected with or without a wire?

你更偏爱通过有线还是无线连接?

How can mother trees take care of seedlings far away?

母树怎样才能照顾远处的幼苗呢?

Do you like the idea that companies know everything about you?

公司能掌握你的所有信息,对此你会介意吗?

Do It Yourself!

自己动手！

Let's find out what people imagine the internet of the future to be like. Start by asking others if they can think of other ways to get data and exchange information, be it with or without any wires. The internet of the fungus, for one, requires everything to be connected through a fine mesh of wires. What other alternatives are there? Look into the options: wires, wireless, and by making use of light. Familiarise yourself with each of the options. Draw up a proposal that combines the different options, so that we can have the best. Compare the best that you have come up with what your friends and family members have come up with.

我们来探讨一下人们对未来互联网的想象。首先问问其他人是否想过用其他方式来获取数据和交换信息，是通过无线还是有线。菌菇网络要求一切都通过细密的线来连接。还有什么其他选择？有线，无线，或者利用光。熟悉每个选项的情况。拟定一个方案，将不同的选项结合起来，以便我们能有

最好的选择。将你想出的最好方案与亲朋好友的进行比较。

学科知识
Academic Knowledge

生物学	丛枝菌根；外生菌根林；蘑菇估计有150万种；森林土壤中90%的生物量是菌菇；菌丝体与子实体之间的差异；蘑菇的不同代谢方式，它们从不摄食，但会在其体外消化和代谢。
化 学	土曲霉制成的衣康酸被用于玻璃和珠宝行业；菌菇会使木材腐烂，并向大气增加800亿吨二氧化碳；菌菇会产生抗生素来保护食物源。
物 理	菌菇降解硬木需要50至100年。
工程学	"最后一英里"的挑战在于铺设的光纤不直接连接到房屋，也不直接连接屋内设备（有线电视除外），因此需要有无线连接。
经济学	食用菌有5 000种，已产业化种植的有200种。蘑菇农场的资本投资极少，可以用极少的资源开展经营；中国拥有世界上最大的"蘑菇经济"，其次是意大利。
伦理学	背信行为；花时间学习和理解另一种语言。
历 史	菌菇是最早被驯化的野生物种，25 000年前最初用来制作面包、酸奶、啤酒和葡萄酒；1845年，霉菌导致土豆腐烂，在爱尔兰造成了大饥荒，并引发了移民北美洲的浪潮；菌菇一直被归类为植物，直到1969年才被单独划分出来；1997年，具有6 000个基因的面包酵母的基因组被发现。
地 理	1980年，美国华盛顿州圣海伦火山爆发后，第一个重新生长出来的生物是蘑菇；蜂蜜蕈在美国俄勒冈州马勒国家森林占有3 726 563平方米的面积，估计已有8 650年的历史。
数 学	从网上冲浪中抽取消费者行为数据；当前和曾经的购物及浏览偏好数据，范围从交易数据到网站流量甚至是社交媒体帖子，能通过预测算法，用于推断未来可能发生的情况。
生活方式	空调与密闭建筑结合本来是为了提高能效，却为菌菇创造了繁殖空间，导致建筑物病态综合征；面包即使在低温下也会发霉；通过"我为人人，人人为我"的默契达成社会凝聚力。
社会学	灵芝作为传统中医药的基石，已有4 500多年的历史；西方文化引发了人们对菌菇的恐惧，尽管只有极少种菌菇是有毒的。
心理学	我们更倾向于对不喜欢的事情视而不见；拥有好奇心的重要性；背叛的心理学，即知道有背叛存在。
系统论	蘑菇会使动植物腐败，并产生营养物用于新用途；树木和蘑菇用碳交换糖。菌菇可确保森林中营养的平衡。

情感智慧
Emotional Intelligence

年轻的甲虫

年轻的甲虫流露出沮丧。他不理解，又十分好奇。他思考了各种可能并承认这是个谜。当年长的甲虫暗示是蘑菇时，他立即放弃了这种选项，因为在他看来蘑菇不能和树对话。他的思路较窄，因为他认为交流仅限于使用语言。他面临的事实与他拥有的知识不符。他提出一系列问题，并将信息与个人经历进行比较，结果甲虫不仅发现这种交流确实存在，而且还发现了更多有关母树及其幼苗怎样保持联系并彼此照顾的知识。

年长的甲虫

年长的甲虫在失望之余怀疑有间谍。她想调查，但是年轻甲虫的看法合情合理，即它们都不会讲树木的语言，所以甲虫当中不可能出现问题，她很快将罪魁祸首认定为蘑菇。然后，她告诉年轻的甲虫，蘑菇具有与树木交流的能力。她重申树木与蘑菇不仅有交流，还有合作。她让自己的直觉结论合理化，使之逐渐清晰并利用了逻辑。在此过程中，她与年轻的甲虫分享了见解，并且对菌菇在黑暗的地下工作的能力发出赞赏。但是，这确实引发了更基本的问题——有些人掌握了更多的信息，从而具有更了解别人的能力。

艺术
The Arts

如何形象地表示网络？在纸上绘制，很容易在平面上做到这一点。但网络是三维的。一种方法是拿一些PET塑料瓶，然后稍稍加热。当塑料变软时，取一块木头，例如火柴棒或小树枝，然后拉出细细的塑料丝。这些细线在空气中很快会变硬。用这些线来模拟蘑菇菌丝体网络，火柴棒或小树枝代表树木。小心操作，创建你独特的艺术品。

思维拓展
Systems: Making the Connections

自然界中的交流是了不起的新发现。除了人类，很多其他生命形式也具备交流信息的能力。比如鲸能够远距离交流，甚至使用音节，并且鲸的语言内容丰富，就像人类一样每个地方的语言都有所不同。我们从这则寓言中学到的是，这不仅与语言有关，还与基础设施的建立有关，基础设施应该有利于实现信息交换，并且允许大范围的产品和能量交换。交流有着明确的目的：合作并保护当地生态系统。菌菇网络促进了亲缘关系，并确保交换会优化母树对其后代的联系和支持。网络中有一个明确的目的：创建社区并确保其韧性。

当气候变化导致物种转变，旧的优势树种必须让位给新的树种时，菌菇网络会意识到这种变化，并以很高的效率促进新物种的进入。这种程度的智能网络还是全新的，并不存在于人类范畴中。正是这种看法让我们能够畅想未来的互联网，用以指导社区所有成员变得更具韧性和适应性。互联网将在未来几年或者几十年中发生转变。菌菇创建的网络既有效率还有目标，并且显然旨在支持"公共事业"。这为我们提供了想象未来互联网的方向。毫不奇怪，这个局域网同时也是交易平台，提高了社区所有成员的效率。当我们发现"母"树可以把碳传送到需要的幼苗时，我们才会意识到这不仅仅只是菌菇网络，而是一个经过几千年进化而来的网络——生命之网。

动手能力
Capacity to Implement

重新思考网络很有必要的。大量研究表明越来越多的人依赖持续的信息交流所提供的动力。所以基于这则寓言，让我们重新设计一种我们所乐见其成效的网络。如何让在大自然中已实现的设计成为我们旨在公益服务的互联网设计的蓝图？

故事灵感来自
This Fable Is Inspired by

苏珊娜·西马德
Suzanne Simard

苏珊娜·西马德是加拿大不列颠哥伦比亚省一个伐木家庭的女儿。她是不列颠哥伦比亚大学森林与保护科学系的森林生态学教授。她的研究聚焦如何理解植物、微生物、土壤、碳、养分和水之间的重要关系，这些关系是生态系统的适应性、复原力和恢复的基础。她主要研究森林以及草原、湿地、苔原和高山生态系统。她对连接森林中生物的地下生态网络的研究很有名，这些网络是生态系统所具有的复杂适应性的基础。

图书在版编目（CIP）数据

冈特生态童书.第七辑:全36册:汉英对照/
(比)冈特·鲍利著;(哥伦)凯瑟琳娜·巴赫绘;
何家振等译.—上海:上海远东出版社,2020
ISBN 978-7-5476-1671-0

Ⅰ.①冈… Ⅱ.①冈… ②凯… ③何… Ⅲ.①生态
环境–环境保护–儿童读物—汉英 Ⅳ.①X171.1-49

中国版本图书馆CIP数据核字(2020)第236911号

策　　划　张　蓉
责任编辑　程云琦
封面设计　魏　来李　廉

冈特生态童书

菌菇网

[比]冈特·鲍利　著
[哥伦]凯瑟琳娜·巴赫　绘

朱　溪　译

记得要和身边的小朋友分享环保知识哦！
八喜冰淇淋祝你成为环保小使者！